THE POETRY OF SCANDIUM

The Poetry of Scandium

Walter the Educator™

SKB
Silent King Books a WhichHead Imprint

Copyright © 2023 by Walter the Educator™

All rights reserved. No part of this book may be reproduced in any manner whatsoever without written permission except in the case of brief quotations embodied in critical articles and reviews.

First Printing, 2023

Disclaimer
This book is a literary work; poems are not about specific persons, locations, situations, and/or circumstances unless mentioned in a historical context. This book is for entertainment and informational purposes only. The author and publisher offer this information without warranties expressed or implied. No matter the grounds, neither the author nor the publisher will be accountable for any losses, injuries, or other damages caused by the reader's use of this book. The use of this book acknowledges an understanding and acceptance of this disclaimer.

"Earning a degree in chemistry changed my life!"
– Walter the Educator

dedicated to all the chemistry lovers, like myself, across the world

CONTENTS

Dedication v

One - Precious And Rare 1

Two - Grand Tale Of Elements 3

Three - Scandium, Oh Scandium 5

Four - Wonders Of Nature 7

Five - Timeless Dance 9

Six - Secrets Deep 11

Seven - Futures Bold 13

Eight - Boundless Flight 15

Nine - Luminescent Lure 17

Ten - Enigmatic And Pure 19

Eleven - Unique Array 21

Twelve - Silent Strength 23

Thirteen - Shimmering Light 24

Fourteen - Path To The New	. . . . .	26
Fifteen - Relentless March	. . . . .	28
Sixteen - Spark Of Creation	. . . . .	30
Seventeen - Soaring Up High	. . . . .	32
Eighteen - Spirits Free	. . . . .	34
Nineteen - Crucible Of Progress	. . . . .	36
Twenty - Symbol Of Might	. . . . .	38
Twenty-One - Bold And Bold	. . . . .	40
Twenty-Two - Dreams Coincide	. . . . .	42
Twenty-Three - Gratitude And Love	. . . . .	44
Twenty-Four - Potential Unbound	. . . . .	46
Twenty-Five - Aerospace Dreams	. . . . .	48
Twenty-Six - Radiant Glow	. . . . .	50
Twenty-Seven - Progress And Wonder	. . . .	52
Twenty-Eight - Celestial Height	. . . . .	54
Twenty-Nine - Metallic Form	. . . . .	56
Thirty - Rare And Bright	. . . . .	57
Thirty-One - Dazzling Light	. . . . .	59
Thirty-Two - Wings Of Innovation	. . . . .	61
Thirty-Three - Cosmic Harmony	. . . . .	63

Thirty-Four - Endless Love	65
Thirty-Five - Catalyst For Change	67
Thirty-Six - Ethereal Glance	69
Why I Created This Book?	71
About The Author	72

ONE

PRECIOUS AND RARE

In the depths of the earth, you quietly reside,
A treasure of nature, so pure and refined.
Scandium, you shimmer with silver-white light,
A rare element, a marvel in sight.

Your atomic number, twenty-one so fair,
Your presence elusive, beyond compare.
In minerals and ores, you hide from the eye,
Yet your beauty and strength, none can deny.

Alloyed with aluminum, you lend them your might,
Creating materials both strong and light.
A catalyst, you are, in the world of science,
Unveiling reactions with quiet compliance.

In aerospace and industry, you play a key role,
Empowering progress, reaching for the goal.
Oh, Scandium, element of wonder and grace,
In the periodic table, you find your place.

So here's to you, Scandium, precious and rare,
A silent force, beyond compare.
May your secrets unravel, may your story unfold,
In the symphony of elements, your tale will be told.

TWO

GRAND TALE OF ELEMENTS

In the earth's crust, you quietly reside,
Scandium, a treasure, hard to find,
Your rarity makes you a gem so pure,
A shining beacon, strong and sure.

Light as a feather, yet tough as steel,
You make alloys that break the wheel,
In aerospace, you soar high and free,
Empowering progress for all to see.

Catalyst of change in reactions bold,
Your presence sparks wonders untold,
In laboratories, your magic unfolds,
Unraveling mysteries, as time foretold.

Industries marvel at your mighty grace,
As you lend strength to every space,

In the periodic table, you stand apart,
A silent legend, waiting to impart.
    Oh Scandium, reveal your secrets untold,
Let your story be written, your history unfold,
In the grand tale of elements, you hold a key,
To unlock the wonders of the world, for all to see.

THREE

SCANDIUM, OH SCANDIUM

In the heart of Earth's crust, a hidden gem,
Scandium, rare and precious, a secret in its realm.
A metal of strength, yet light as a dream,
Unveiling its mysteries, a celestial theme.

In aerospace it soars, a silent companion,
Forged in the fires of stars, a celestial union.
Alloyed with aluminum, a bond so true,
Creating wings of wonder, the skies it'll pursue.

In the factories it toils, a worker of might,
Enhancing the tools, in the glow of the light.
A catalyst for change, in the chemistry of life,
Unveiling its power, amidst the daily strife.

Oh Scandium, enigmatic and bold,
Revealing your essence, a tale untold.
In the grand tapestry of elements, you hold your

place,
A symbol of wonder, in the cosmic embrace.
 So let us unravel your secrets, one by one,
For in your depths lie treasures, yet to be spun.
A metal of marvel, a beauty concealed,
Scandium, oh Scandium, to you we shall yield.

FOUR

WONDERS OF NATURE

In the earth's crust, a hidden gem does lie,
Scandium, rare and precious, catching the eye.
With strength and grace, it stands apart,
A metal of wonder, a work of art.
   From aerospace to industry's embrace,
Scandium's prowess finds its place.
Light and strong, it takes to the sky,
Aiding machines as they soar and fly.
   Origin unknown, its secrets held tight,
A mystery element, a celestial light.
Enhancing tools with a magical touch,
Catalyzing change, it does so much.
   Oh, Scandium, enigmatic and rare,
In the grand tapestry, you have your share.

Untold treasures within your core,
We long to know you, to explore more.
    So, let us unravel your mysteries deep,
And unlock the secrets you lovingly keep.
For in your essence, there lies a key,
To the wonders of nature, for all to see.

FIVE

TIMELESS DANCE

In the earth's crust, you quietly lie,
Scandium, a treasure, hidden from the eye.
Rare and precious, your presence is slight,
Yet your strength and beauty shine so bright.
    Light as a feather, yet strong and true,
You whisper secrets of the skies so blue.
In aerospace and industry, you play your part,
A metal of wonder, with a place in every heart.
    Catalyzing change, you work behind the scenes,
Enhancing tools, fulfilling dreams.
A silent force, with power untold,
In the grand tapestry of elements, your story unfolds.
    Oh Scandium, element of grace,
In your quiet elegance, we find our place.

Mysterious and rare, yet steadfast and strong,
In your timeless dance, we all belong.

SIX

SECRETS DEEP

In the earth's embrace, Scandium lies,
A rare gem hidden from prying eyes,
Forged in stars, a celestial dance,
Its essence weaves a mystic trance.

    A metal of strength, yet light as air,
It soars through space without a care,
In aerospace it finds its home,
Where dreams take flight and stars are sewn.

    Its secrets whispered in the wind,
A catalyst for change, it's been pinned,
Unraveling mysteries, forging new ways,
In its enigmatic dance, it sways.

    A shimmering beauty, a silent force,
In the grand tapestry, it holds its course,

Amidst the elements, it stands alone,
A testament to the unknown.
    Oh Scandium, with your secrets deep,
In your silent wisdom, may we keep,
A glimpse of truth, a touch of grace,
In your cosmic dance, we find our place.

SEVEN

FUTURES BOLD

In the heart of stars, Scandium gleams,
A whispering presence in cosmic streams.
Light as a feather, yet strong and true,
Unveiling secrets, old and new.
    A catalyst of change, it dances in flight,
Unraveling mysteries, embracing the light.
In aerospace realms, it yearns to soar,
Opening doorways to worlds unknown.
    Enigmatic Scandium, a marvel untold,
In the earth's embrace, a treasure to behold.
Its atomic dance, a symphony rare,
Weaving dreams with delicate care.
    Like a silent guardian, it stands the test,
In alloys and structures, it gives its best.

Majestic and rare, a shimmering grace,
Unveiling the future, in its silent embrace.
    Oh Scandium, enigma of the Earth and sky,
In your timeless dance, we can't help but sigh.
For in your essence, a promise untold,
Of mysteries unlocked and futures bold.

EIGHT

BOUNDLESS FLIGHT

    In the heart of stars, Scandium glows bright,
A silent dancer in the cosmic night,
Light as a feather, yet strong and bold,
Unraveling secrets, a story untold.
    Rare and enigmatic, a jewel in the earth,
Whispering promises of a new rebirth,
Catalyst of change, in aerospace it soars,
Defying gravity, unlocking closed doors.
    Magnetic allure, a mesmerizing grace,
Invisible threads, weaving through space,
A symphony of atoms, a delicate song,
Scandium's melody, pure and strong.
    Oh, elusive element, with powers unseen,
In your essence, lies a future serene,
Industry's ally, a lustrous dream,
Shaping tomorrow, with a radiant gleam.

So, let's raise a toast to Scandium's might,
A shimmering beacon, in the realm of light,
May its mysteries unfold, in wondrous delight,
Guiding humanity, towards boundless flight.

NINE

LUMINESCENT LURE

In the heart of the earth, a hidden treasure lies,
Scandium, a marvel with enigmatic eyes.
A whisper in the wind, a secret untold,
It sparkles with strength, a story to unfold.
    Light as a feather, yet mighty and bold,
In aerospace dreams, its wonders are told.
An element rare, a shimmering dance,
Unraveling mysteries with each fleeting glance.
    Catalyst of change, in the alchemy of time,
Unlocking the secrets, a rhythm so sublime.
A symphony of atoms, a celestial song,
Scandium's allure echoes strong and long.
    With grace and resilience, it weaves a tale,
Of industry's prowess, it will never fail.

Soaring to heights, where dreams take flight,
Scandium, a beacon, in the darkest night.
    Oh, Scandium, your essence so pure,
In the tapestry of elements, a luminescent lure.
A silent guardian, a celestial key,
Revealing the universe's wondrous mysteries.

TEN

ENIGMATIC AND PURE

In the heart of stars, Scandium lies,
A silent force, a hidden prize.
With strength concealed, yet bold and bright,
It sparks the flames of boundless flight.
    In aerospace dreams, it takes to the sky,
Unleashing potential, soaring high.
Catalyst of change, it whispers its call,
Unveiling the future, breaking down walls.
    A lustrous gem in the earth's embrace,
Unyielding, resilient, with quiet grace.
Unraveling secrets, with a touch so light,
It dances with atoms, in the depths of night.
    Oh Scandium, enigmatic and pure,
In industry's grasp, you endure.

Forging alloys of strength and might,
Shaping tomorrow, in the forge of light.
    Mysterious element, with tales untold,
In your essence, wonders unfold.
Soaring through space, embracing the new,
Oh Scandium, we marvel at you.

ELEVEN

UNIQUE ARRAY

In the heart of stars, Scandium gleams,
A glint of strength, beyond earthly dreams.
In aerospace, it soars and flies,
Unleashing power, beyond the skies.

Its atomic dance, a wondrous sight,
Catalyzing change, with sheer delight.
A metal of grace, in rare supply,
Unraveling mysteries, up so high.

From engines roaring, to frames so light,
Scandium whispers, through the endless night.
A shimmering element, bold and rare,
Shaping the future, with tender care.

In laboratories, its secrets unfold,
Unlocking potential, in stories untold.

A silent hero, in the periodic play,
Guiding innovation, in a unique array.
    Oh Scandium, with beauty so true,
We marvel at the wonders, that spring from you.
In every atom, a world to explore,
A radiant allure, forevermore.

TWELVE

SILENT STRENGTH

In the heart of stars, a silent dance begins,
Where atoms weave a symphony to light.
Scandium, the enigmatic metal spins,
Guiding humanity towards boundless flight.

A shimmering beacon, a whispering call,
Unraveling mysteries in its embrace.
Strength and grace, within its atoms, enthrall,
As it shapes the future, leaving no trace.

A catalyst for change, it takes its place,
In aerospace realms, where dreams take to flight.
Melding with alloys, it adds boundless grace,
Helping humans reach new heights, bold and bright.

Oh Scandium, your secrets we unfold,
In your silent strength, our story's told.
A hero in the realm of elements,
Guiding us to the universe's expanse.

THIRTEEN

SHIMMERING LIGHT

In the heart of stars, Scandium's dance begins,
A metal of wonder, beyond earthly sins.
In aerospace dreams, it takes to the sky,
Unlocks mysteries, as it soars up high.

Its allure magnetic, drawing the eye,
In alloys and engines, it's destined to fly.
A whisper of change, in the air it weaves,
Guiding humanity to boundless eaves.

With strength and grace, it shapes our machine,
A catalyst for progress, yet to be seen.
In laboratories it sparks innovation's flame,
Leading us forward, without any blame.

Oh Scandium, element of the future's hold,
In your shimmering light, stories unfold.

A promise of flight, to the stars we aspire,
With you as our guide, we'll reach ever higher.

FOURTEEN

PATH TO THE NEW

In the heart of stars, Scandium gleams,
A catalyst for dreams, beyond earthly streams.
Its magnetic allure, a force so pure,
Guiding us to soar, to futures obscure.

A whisper in the wind, a dance in the sky,
Scandium's grace, never asking why.
In aerospace it thrives, a silent guide,
Unveiling mysteries, where dreams collide.

In engines it sings, a symphony of might,
Propelling us forward, through boundless flight.
In alloys it weaves, strength refined,
Shaping our world, a destiny enshrined.

A harbinger of change, a beacon so bright,
Scandium's resilience, a relentless light.
Unlocking the secrets, of ages untold,
A symphony of atoms, a story to unfold.

Oh, Scandium, noble and true,
In your embrace, the future brews.
A lighthouse in the storm, a path to the new,
In your essence, humanity finds its due.

FIFTEEN

RELENTLESS MARCH

In the heart of stars, Scandium gleams,
A catalyst for dreams and schemes.
Unlocking mysteries, shaping the future bright,
It dances with potential, a guiding light.

A metal rare, yet mighty in its ways,
It sparks innovation, igniting the blaze.
In aerospace dreams, it takes flight,
Guiding humanity towards boundless height.

With strength and grace, it paves the way,
In engines and alloys, it holds sway.
A whisper of wonder, a touch of allure,
Scandium beckons, promising so much more.

In industry's embrace, it stands tall,
A silent giant, commanding all.

With power to transform, it leads the charge,
In the ever-evolving, relentless march.
　So raise a toast to Scandium's might,
As it propels us forward, into the night.
A hero unsung, yet steadfast and true,
Shaping our world, with a shimmering view.

SIXTEEN

SPARK OF CREATION

In the light of stars, Scandium gleams,
A silent catalyst of our aerospace dreams.
Hidden in ores, a treasure untold,
Unleashing potential, a story unfolds.

Mysterious element, graceful and strong,
Innovating the future, where we belong.
With strength and allure, it takes to the sky,
Guiding us forward, as we learn to fly.

A spark of creation, in its atomic dance,
Unraveling secrets, with each cosmic chance.
From Earth to the heavens, it paves the new way,
In Scandium's embrace, we find our display.

Soaring through limits, breaking through bounds,
Scandium, the muse, where innovation resounds.
In alloys and structures, it shapes the unknown,
A symphony of progress, in its elemental tone.

Oh, Scandium, element of grace and might,
Igniting the spark, in the depths of the night.
A beacon of hope, in the scientist's hand,
Unveiling the future, where dreams expand.

SEVENTEEN

SOARING UP HIGH

In the heart of stars, where fusion dances bright,
Scandium emerges, a beacon of light.
A catalyst for change, it sparks innovation's flame,
Guiding us forward, never staying the same.

In aerospace dreams, it takes to the sky,
Unleashing potential, soaring up high.
With strength and allure, it shapes what's to come,
Innovating our world, never feeling numb.

Boundless flight beckons, with Scandium's embrace,
Unlocking mysteries, transcending time and space.
A metal of wonder, a shimmering gleam,
Innovating our world, like a beautiful dream.

So let's raise our voices, in awe and in praise,
For Scandium's power, in countless new ways.

Catalyzing change, it leads us to find,
A future where possibilities endlessly bind.

EIGHTEEN

SPIRITS FREE

In the heart of engines and wings it lies,
Scandium, a marvel that defies the skies.
A metal rare, with strength untold,
In aerospace dreams, its story unfolds.
　　Alloyed with aluminum, it takes flight,
Defying gravity with all its might.
In engines roaring, it plays a part,
Pushing boundaries, igniting the heart.
　　Mysteries unlocked, by its shimmering glow,
In laboratories, where innovations grow.
A catalyst for change, it leads the way,
Guiding humanity to a brighter day.
　　Oh, Scandium, your allure so grand,
In the palm of our hands, you boldly stand.
With each discovery, a new chapter unfurls,
As we harness your power to shape the world.

So, let us marvel at your atomic dance,
And embrace the future, with every chance.
For in your essence, we find the key,
To reach boundless heights, and set our spirits free.

NINETEEN

CRUCIBLE OF PROGRESS

In the heart of engines and wings, you dwell,
Scandium, element of strength and grace,
Aerospace marvel, in your secrets, we delve,
Innovations sparked by your radiant trace.

Resilient metal, forged in the stars,
Unyielding ally in the quest for flight,
Your presence propels us past earthly bars,
To soar unfettered, in the boundless height.

Mysterious Scandium, unlocker of doors,
Revealing the secrets of alloys untold,
With your shimmering presence, the future soars,
In your embrace, innovation takes hold.

In the crucible of progress, you stand tall,
Shaping the world with your potent might,

Guiding humanity, as we heed your call,
Towards a future, bathed in your luminous light.

TWENTY

SYMBOL OF MIGHT

In the heart of stars, you were born,
Scandium, element of grace and might,
A silent force in the fabric of space,
Unraveling mysteries, shaping the flight.

In aerospace dreams, you take your place,
Light and strong, defying the earth's pull,
Lifting us high, beyond the clouds,
Where boundaries fade, and dreams are full.

With strength unmatched, you stand tall,
A catalyst for change, an alloy of power,
Unleashing innovation, igniting the heart,
In the crucible of progress, you are the hour.

Your allure is undeniable, your potential vast,
Unlocking new frontiers, pushing the boundaries,

Scandium, element of wonder and might,
In your essence, humanity finds its stories.
    So here's to you, Scandium, shining bright,
In the tapestry of life, you weave your art,
A metal of dreams, a symbol of might,
Forever shaping the world, forever a part.

TWENTY-ONE

BOLD AND BOLD

In the forge of aerospace, you reign supreme,
Scandium, metal of strength and dream.
Your allure, like a siren's call, pulls us high,
Defying gravity, reaching for the sky.
  With strength unmatched, you shape the world,
In alloys and frames, your power unfurled.
Guiding the flight, with potent might,
Innovating the future, shining so bright.
  Catalyst for change, unlocking mysteries deep,
Scandium, you stir the secrets we keep.
Leading humanity towards a path unseen,
In your shimmering essence, we find the sheen.
  A touch of magic, in your atomic dance,
You hold the power of progress, in every glance.

Defying the odds, with resilience untold,
Scandium, you shape the future, bold and bold.

TWENTY-TWO

DREAMS COINCIDE

In the heart of the earth, Scandium lies,
A metal of wonder that defies the skies.
Unseen and unnoticed, yet mighty and true,
It holds the power to change our view.

    Aerospace dreams take flight with its grace,
Unveiling mysteries, pushing boundaries in space.
Light as a feather, yet strong as can be,
It unlocks the heavens for all to see.

    With strength and allure, it shapes the world,
In ways unfathomed, its potential unfurled.
Defying gravity, it soars to new heights,
Guiding humanity towards radiant lights.

    Catalyst for change, it sparks the flame,
Igniting innovation, never the same.

Secrets untold, within its core,
Scandium whispers of worlds to explore.
    Oh, noble element, we honor thee,
For all that you are, and all that will be.
In your shimmering essence, we find our guide,
Towards a future where dreams coincide.

TWENTY-THREE

GRATITUDE AND LOVE

In the dance of elements, Scandium shines bright,
Defying gravity, it soars through the night.
Unlocking the heavens with its radiant glow,
Guiding humanity where dreams tend to flow.

A catalyst for change, innovation's true friend,
Whispering of secrets, worlds without end.
Its shimmering essence, a beacon so pure,
Leading us onward, our hopes to secure.

Unraveling mysteries, untold and unseen,
Scandium's touch paints the world pristine.
A metal so rare, yet with power untold,
Shaping the future, a story to unfold.

In laboratories it sparks, a scientist's delight,
In spacecraft it journeys, through the stars it takes flight.

A marvel of nature, a gift from above,
Scandium, we honor you, with gratitude and love.
    So here's to the element that lights up the way,
Guiding us forward, where dreams coincide and play.
In the tapestry of life, you shimmer and gleam,
Scandium, oh Scandium, our beacon, our dream.

TWENTY-FOUR

POTENTIAL UNBOUND

In the heart of innovation, Scandium gleams,
A metal of dreams, defying earthly regimes.
In aerospace realms, it soars with might,
Guiding humanity towards boundless flight.

A shimmering essence, rare and pure,
Unlocks mysteries, a catalyst sure.
Pushing boundaries, setting spirits free,
In its radiant allure, we find the key.

Resilient and strong, it shapes the world,
A beacon of hope, its power unfurled.
Mighty Scandium, we honor your grace,
As you lead us to a brighter place.

So we raise our voices, in gratitude sing,
For the dreams you ignite, on majestic wing.

In laboratories and skies, your potential unbound,
Scandium, metal of wonder, where dreams are found.

TWENTY-FIVE

AEROSPACE DREAMS

In the heart of stars, a secret lies,
Scandium, a metal that defies,
Gravity's pull, it reaches high,
Unlocking heavens, beyond the sky.

A catalyst for dreams untold,
Innovation's story, yet to unfold,
Strength and allure, in every glance,
Scandium leads the cosmic dance.

A metal rare, yet bold and bright,
Shaping the future, with guiding light,
In aerospace dreams, it takes its place,
Pushing boundaries, with boundless grace.

O Scandium, we sing to thee,
For all the wonders, yet to be,
In laboratories, and in space,
You lead us to an unknown place.

So here's to Scandium, noble and true,
We cherish the dreams, you help us pursue,
A beacon of hope, in a world so vast,
Scandium, metal of dreams, forever will last.

TWENTY-SIX

RADIANT GLOW

In the boundless expanse, Scandium soars,
A metal of marvel, unlocking cosmic doors.
Aerospace dreams, it fervently ignites,
Pushing the limits, reaching celestial heights.

In engines it dances, a shimmering light,
Unraveling mysteries, taking flight.
With strength and allure, it conquers the skies,
Unleashing potential, where innovation lies.

In laboratories, it whispers of change,
A catalyst for progress, an element so strange.
Shaping the future with its radiant glow,
Scandium, the trailblazer, continues to grow.

Oh, Scandium, metal of heavenly grace,
With your shimmering essence, you boldly embrace

The challenges posed by the cosmos above,
Unlocking the heavens with strength and love.

TWENTY-SEVEN

PROGRESS AND WONDER

In the realm of aerospace dreams, you gleam,
Scandium, the key to unlock the heavens high.
With strength and grace, you pierce the sky,
Your beauty and allure, a celestial theme.

Innovative spirit, catalyst for change,
You dance with atoms, rearrange,
Transforming the mundane into something new,
Oh Scandium, your potential, a wondrous view.

A metal of wonder, a metal of might,
Shaping the world with your radiant light.
Guiding humanity towards a brighter tomorrow,
In your presence, we find solace in our sorrow.

Oh Scandium, with strength unknown,
You sculpt the future from your throne.

A shimmering beacon, a promise untold,
In your embrace, mysteries unfold.
    Pushing the limits, reaching celestial heights,
In aerospace and laboratories, you ignite.
Your essence, a symphony of progress and wonder,
Oh Scandium, in you, we find the world to ponder.

TWENTY-EIGHT

CELESTIAL HEIGHT

In the heart of stars, Scandium glows,
A shimmering essence, a mystery it bestows.
Unlocking the heavens, shaping the future's design,
In the cosmic dance, its allure will shine.
    A metal of dreams, aerospace's delight,
Pushing boundaries, reaching celestial height.
Mingling with aluminum, forging strength untold,
In the boundless sky, its story will unfold.
    A whisper in alloys, enhancing the light,
An element of wonder, a beacon in the night.
With each atomic dance, its potential takes flight,
In the grand cosmic theater, it weaves its plight.
    So let us marvel at Scandium's grace,
In the tapestry of elements, it finds its place.

Unlocking the mysteries, shaping the world's decree,
In the grand cosmic symphony, it dances free.

TWENTY-NINE

METALLIC FORM

In the starry expanse, Scandium shines bright,
A metal of wonder, a beacon of light.
In aerospace it soars, defying the earth,
Unlocks the mysteries, of the cosmos' birth.

Its atomic dance, a marvel to behold,
Unleashing potential, yet untold.
Innovation's ally, it takes to the skies,
Pushing the boundaries, where limits arise.

A catalyst for change, it sparks the flame,
Transforming the mundane, into something untamed.
With strength and allure, it captures the eye,
Shaping the future, where dreams reach the sky.

Oh Scandium, beauty in metallic form,
In your shimmering essence, we find a new norm.
Unlocking the heavens, in cosmic dance,
Revealing the secrets, of fate's grand romance.

THIRTY

RARE AND BRIGHT

In realms beyond our earthly bounds,
Where stars ignite and worlds abound,
There lies a metal, rare and bright,
Scandium, a beacon in the night.
    Its glow reveals the cosmic dance,
A key to realms beyond mere chance,
In laboratories, it takes its place,
Unraveling the mysteries of space.
    With strength and grace, it takes to flight,
In wings of planes, a shining light,
Pushing boundaries, reaching high,
In aerospace, it seeks to fly.
    A metal of the future, bold and true,
Unleashing potential, not yet through,

In engines roaring, and structures strong,
It shapes the world, where it belongs.
    So let us marvel at this wondrous element,
Its allure, its beauty, so resplendent,
For in its power, we find the key,
To unlock the heavens, and set our spirits free.

THIRTY-ONE

DAZZLING LIGHT

In the heart of the cosmos, Scandium shines,
A celestial dancer, in rare, mystic lines.
With strength and allure, it breaks through the night,
Unlocks the heavens, in its shimmering light.
   In laboratories, it weaves its own tale,
A catalyst for change, it will never fail.
Pushing boundaries, shaping the future's design,
Revealing the mysteries, of the cosmic divine.
   Aerospace marvel, soaring high and free,
Scandium whispers secrets, of what's yet to be.
In alloys it binds, with metals it blends,
A beacon of hope, as the universe extends.
   Oh, Scandium, element of grace,
In your atomic dance, we find our place.

With potential untold, you guide us to see,
The wonders of creation, in all their beauty.
    So here's to Scandium, so rare and so bright,
A key to the heavens, a dazzling light.
In your cosmic embrace, we find our way,
To explore, to discover, to dream and to stay.

THIRTY-TWO

WINGS OF INNOVATION

In the heart of stars, a secret lies,
A metal rare, that dares to rise.
Scandium, oh Scandium, so pure and bright,
In your atomic dance, you paint the night.

A catalyst for dreams, you soar on high,
Unveiling mysteries in the endless sky.
With strength untold, you break the mold,
Pushing boundaries, a story to be told.

In aerospace realms, you find your place,
A metal of wonder, a beacon of grace.
With wings of innovation, you take flight,
Guiding us through the boundless night.

In laboratories, you work your charm,
Revealing the cosmos, keeping us warm.

Your allure, like a celestial song,
Inspires our hearts, where we belong.
   Oh Scandium, you hold the key,
To unlock the heavens, set our spirits free.
A metal of dreams, a shimmering light,
Guiding us forward, through day and night.

THIRTY-THREE

COSMIC HARMONY

In the heart of stars, Scandium gleams,
A celestial dance in cosmic streams,
A beacon of light, a symphony's grace,
Unveiling the secrets of time and space.
    Innovator of flight, aerospace's muse,
Elevating dreams, where limits diffuse,
With strength and resilience, its beauty unfurls,
Guiding humanity to explore the world.
    A shimmering metal, a vision of might,
Unveiling the cosmos, in boundless flight,
Unlocking the heavens, with each atomic trace,
Scandium, the catalyst of cosmic embrace.
    In the dance of creation, it takes center stage,
A catalyst for change, in every age,

Pushing boundaries, shaping our destiny,
Scandium, the element of cosmic harmony.

THIRTY-FOUR

ENDLESS LOVE

In the forge of stars, Scandium was born,
A shimmering ember, a celestial adorn.
With strength untold and grace divine,
It lights the path where dreams entwine.
    From aerospace skies to the depths below,
Scandium's allure, a wondrous show.
Catalyst of change, it whispers in the wind,
Unraveling mysteries, where hopes ascend.
    In alloys it dances, a symphony of might,
Forging the future, in the cosmic light.
A beacon of innovation, it leads the way,
To realms unexplored, where visions sway.
    Oh, Scandium, key to the heavens above,
Unleash the boundless, the endless love.

Guide humanity's quest, in time and space,
Reveal the secrets, of this wondrous place.
　　Soar, Scandium, on wings of desire,
Ignite the sparks, of celestial fire.
In your essence, we find our dreams,
As you unravel the cosmos, in radiant beams.

THIRTY-FIVE

CATALYST FOR CHANGE

In the heart of the earth, Scandium lies,
A metal of wonder that captures our eyes.
With strength and allure, it beckons us near,
Unveiling the secrets we hold dear.
    Aerospace dreams, it helps us to soar,
Unlocking the heavens, forever to explore.
Its shimmering beauty, a sight to behold,
A beacon of progress, a story untold.
    In the depths of creation, it plays a key role,
A catalyst for change, it shapes the whole.
Mysteries of cosmos, it begins to unbind,
Revealing the truths that were once confined.
    With each passing moment, its potential unfolds,
Pushing the boundaries, where destiny molds.

So let's gaze at Scandium, with wonder and awe,
For it holds the future, in its shimmering draw.

THIRTY-SIX

ETHEREAL GLANCE

In the heart of stars, you were born,
Scandium, element of celestial morn.
A shimmering light in the cosmic dance,
Unveiling secrets with each ethereal glance.

With strength and grace, you take to flight,
Aerospace dreams you bring to light.
Alloyed with aluminum, you soar on high,
Pushing boundaries, reaching for the sky.

A catalyst for change, you pave the way,
Innovations bloom where you hold sway.
From bikes to baseball, a touch of your hand,
Unleashes potential across the land.

Mysterious element, rare and bright,
You fuel the fire of human delight.

In laboratories and minds, you spark the flame,
Guiding us toward the unknown, untamed.
    Oh, Scandium, beacon of hope and might,
In your essence, we find our flight.
To the cosmos, we turn our gaze,
And in your story, our destiny lays.

WHY I CREATED THIS BOOK?

Creating a poetry book about the chemical element Scandium was a unique and fascinating way to explore the element's properties, history, and significance. Poetry has the power to convey emotions, tell stories, and capture the essence of a subject. By crafting poems about Scandium, I can delve into its atomic structure, uses in various industries, and even its symbolic representation. This innovative approach can make scientific concepts more accessible and engaging to a wider audience, sparking curiosity and appreciation for the world of chemistry.

ABOUT THE AUTHOR

Walter the Educator is one of the pseudonyms for Walter Anderson. Formally educated in Chemistry, Business, and Education, he is an educator, an author, a diverse entrepreneur, and he is the son of a disabled war veteran. "Walter the Educator" shares his time between educating and creating. He holds interests and owns several creative projects that entertain, enlighten, enhance, and educate, hoping to inspire and motivate you.

Follow, find new works, and stay up to date
with Walter the Educator™
at WaltertheEducator.com

www.ingramcontent.com/pod-product-compliance
Lightning Source LLC
LaVergne TN
LVHW051959060526
838201LV00059B/3742